Beekeeping Guide for Beginners

My Personal Beekeeping Journey

I0465678

By

Luciana Perla

Table of Contents

Preface

Beekeeping is a beautiful, rewarding journey—one that connects you deeply with nature and introduces you to one of the planet's most fascinating creatures: the honeybee. When I first started my own beekeeping adventure, I was filled with excitement, but also a bit of fear and uncertainty. I didn't know if I had the right knowledge, tools, or even the courage to care for a hive of bees. But over time, I learned that beekeeping is less about perfection and more about patience, observation, and respect for these incredible pollinators.

Whether you've already set up your hive or you're just beginning to explore the idea, I hope this guide provides you with the support and understanding you need. Beekeeping is a partnership between you and your bees, and it's a partnership that can transform not only your garden but also your connection to the natural world.

Welcome to the world of beekeeping. I'm so glad you're here, and I'm excited to share this journey with you.

Introduction

Beekeeping is much more than a hobby—it's an invitation to become part of something larger than yourself. When you start keeping bees, you're not just producing honey; you're playing an essential role in supporting biodiversity, nurturing your local ecosystem, and, in many ways, giving back to nature.

For many, the idea of keeping bees can feel intimidating. Perhaps you've heard stories of complicated equipment, swarm management, or even painful bee stings. You might wonder, "Is this something I can really do?" The answer is a resounding yes. Beekeeping may have its challenges, but at its core, it's a practice rooted in simplicity and harmony with nature. Bees, after all, have been managing their own colonies for millions of years. As beekeepers, our role is to assist, observe, and protect—not to control.

The best beekeepers are those who take the time to observe and respect their bees. When you open your hive for the first time, you'll be stepping into a world that operates on cooperation, instinct, and resilience. The honeybees will teach your patience and the importance of balance. In return, you'll help them thrive by providing a safe environment, free from pesticides and other harmful influences.

CHAPTER 1

Beekeeping

Why Start Beekeeping?

If you're reading this, something about the world of bees has caught your attention, and you're likely standing at the threshold of an exciting journey. Maybe it's the idea of harvesting your own honey, the peaceful hum of a thriving hive, or the chance to play a part in supporting our fragile ecosystem that has drawn you in. Whatever your reason, welcome. Beekeeping is not just a hobby—it's a partnership with nature, one that offers beauty, mystery, and fulfillment in every season.

For many, the thought of starting beekeeping can seem overwhelming. There's equipment to understand, seasons to track, and a lot of buzzing creatures to manage. But don't worry—if the idea of tending to bees has sparked something in you, that's enough. You don't need to be an expert right away; all you need is curiosity and patience. Beekeeping has been practiced for thousands of years, from ancient civilizations to backyard hobbyists, and there's a reason why people have been drawn to it for so long.

At its heart, beekeeping is a relationship—a dynamic, evolving connection between you and your bees. It's fascinating to witness the incredible world inside a hive, where each bee has a specific role, and every decision they make contributes to the survival of their colony. Beekeeping lets you observe nature's teamwork up close, and there's something magical about playing a small part in this natural dance.

But beekeeping is more than just observing. It's about understanding that your role is both steward and student. Bees teach us about patience, adaptability, and the interconnectedness of life. They also give us honey, a golden reward that's rich not just in flavor, but in meaning, because it represents months of hard work—from both the bees and you.

The Joys and Challenges: What You Need to Know

Like anything worthwhile, beekeeping comes with its joys and challenges. Let's start with the joys, because there are plenty.

One of the most rewarding parts of beekeeping is the connection it gives you to the natural world. Bees are incredibly sensitive creatures, responding to subtle changes in weather, plant life, and even mood. When you care for bees, you become attuned to these small shifts, noticing the first bloom of spring or the crispness in the air as autumn approaches. You'll start to see the world through the bees' eyes, understanding the delicate balance they maintain with nature.

And then there's the honey. Harvesting honey from your own hive is one of the most satisfying experiences you can have as a beekeeper. It's not just about the sweet taste, though that's a major perk! It's about knowing that this golden nectar is a product of your care and attention. The first time you spread your home-harvested honey on toast or share it with family and friends, you'll feel an immense sense of pride. Every jar of honey carries the story of your journey as a beekeeper.

But with joy also come challenges. Beekeeping requires dedication, patience, and a willingness to learn. You'll face unpredictable weather, pests, and diseases that can threaten the health of your hive. There will be moments when things don't go according to plan—when a colony might struggle, or worse, collapse. These are tough moments, and they can be discouraging.

Yet, these challenges are also where the most valuable lessons lie. Beekeeping teaches you resilience. When faced with setbacks, you learn to observe, adjust, and try again. The bees, with their constant cycles of rebuilding and adapting, show us the importance of perseverance. Every beekeeper has encountered failures, but those who stick with it find that the rewards far outweigh the setbacks.

And remember, you're never alone in this journey. The beekeeping community is vast and supportive. Whether you're connecting with fellow beekeepers online, attending a local beekeeping club, or seeking advice from a mentor, there's always someone willing to help you through the rough patches.

My Personal Beekeeping Journey

Let me take a moment to share a bit about my own beekeeping journey, because like you, I started as a beginner with more questions than answers. I'll never forget the day I decided to bring bees into my life. It was spring, and I'd been reading about the importance of pollinators and the alarming decline in bee populations. I wanted to do something that felt meaningful, something that connected me to nature in a real, tangible way. Beekeeping seemed like the perfect choice.

I bought my first hive, a simple wooden box, and ordered a colony of bees. I'll admit, when the box arrived with thousands of buzzing creatures inside, I was nervous—terrified even. What if I did something wrong? What if the bees didn't like their new home? What if I couldn't handle it?

But as I opened the box and released the bees into their hive, something incredible happened. I realized that beekeeping isn't about control. It's about trust— trusting that the bees know what they're doing and trusting myself to learn alongside them. From that day forward, I've been hooked.

There were plenty of bumps along the way. My first hive didn't make it through the winter, and I remember feeling devastated. I'd spent months tending to them, learning their rhythms, and it felt like a personal failure. But the more I learned, the more I understood that loss is part of beekeeping, just as it's part of life. I tried again the next spring, and this time, my hive thrived.

Beekeeping has taught me patience in ways I never expected. It's made me a better observer of the world, more aware of the seasons and the delicate interplay between plants, weather, and pollinators. It's also given me a deep appreciation for the bees themselves—these tiny, tireless creatures who work together in perfect harmony.

As you begin your own beekeeping journey, know that you will have moments of doubt, and that's okay. You'll make mistakes, and that's okay too. Every beekeeper does. But if you stay curious, patient, and open to learning, the rewards will be more than you ever imagined.

So, welcome to the world of beekeeping. It's an adventure, a partnership, and a chance to connect with nature in a way that is both humbling and awe-inspiring. I'm excited to share this journey with you.

CHAPTER 2

Understanding Bees: The Heart of Your Hive

When you first approach a hive, what you'll notice is the sound—a gentle, steady hum. It's the sound of thousands of bees, each doing their part, working in harmony to keep their colony thriving. As a beginner beekeeper, it's easy to get lost in the mechanics of beekeeping—buying the right equipment, learning how to inspect a hive, harvesting honey—but before you get too far into the practicalities, it's essential to take a step back and understand the bees themselves.

To truly care for your bees, you must understand their world. Every action you take in beekeeping, every decision you make, should be informed by your knowledge of how these incredible creatures live and work together. They are not just insects; they are an organized, highly efficient society. Their lives are intertwined with the environment, and their survival depends on a delicate balance within the hive. Once you understand how they function, you'll appreciate the magic happening right before your eyes.

The Life of a Honeybee: Queen, Worker, Drone

A honeybee colony is a fascinating, organized society with three main types of bees, each playing a vital role in the hive: the queen, the workers, and the drones. Understanding their individual responsibilities will not only help you manage your hive better but also deepen your respect for the intricate system that bees have perfected over millions of years.

At the heart of every hive is the queen, the most important bee in the colony. Though she is only one bee among thousands, her presence is essential. The queen's primary role is to lay eggs—up to 2,000 per day during the height of the season. Without her, the colony would eventually die out. She produces pheromones that serve as a chemical signal to the rest of the hive, regulating the behavior of other bees and maintaining unity.

The queen is the longest-living member of the colony, often surviving several years. But her reign is not necessarily permanent. If she becomes weak, sick, or fails to lay enough eggs, the colony may decide to replace her—a process called "supersedure." The worker bees will raise a new queen, and the two may even duel, with the younger queen often emerging victorious. While this might sound brutal, it's nature's way of ensuring the colony remains strong and healthy.

Worker Bees: The Unsung Heroes

Worker bees are all female, and they are the true backbone of the hive. Everything you see happening in and around the hive is because of their tireless efforts. Workers handle every job, from foraging for nectar and pollen, feeding the queen and larvae, to cleaning the hive, guarding the entrance, and even regulating the hive's temperature. They quite literally work themselves to death, living only about six weeks during the busy summer months.

One of the most fascinating aspects of worker bees is that their role changes as they age. Younger workers stay inside the hive, caring for the queen and her brood (larvae and pupae), while older workers become foragers, flying up to five miles away in search of nectar and pollen. The dedication of worker bees is inspiring, and as a beekeeper, you'll begin to notice their unwavering commitment to the survival of the colony.

Drones: The Gentle Giants

Drones, the male bees, have a singular purpose: to mate with a queen. They are larger than worker bees and do not have stingers, making them completely harmless. Unlike the worker bees, drones don't collect food, care for the young, or do any hive maintenance. Their only role is reproduction, and once they've mated with a queen (which happens mid-flight in a process called a "drone congregation"), they die shortly afterward.

Drones are only present during the warmer months when new queens need to mate. As winter approaches, the worker bees will expel the drones from the hive, as they no longer serve a purpose and would only consume valuable resources. It may seem cruel, but this is another example of the bees' efficiency—everything in the hive exists for a reason, and nothing goes to waste.

Inside the Hive: The Bee's Ecosystem

The hive is more than just a box full of bees; it's a carefully maintained environment where every square inch has a purpose. It's a true ecosystem where each bee knows exactly what needs to be done to keep the colony alive.

The interior of the hive consists of several frames, each holding hexagonal wax cells built by the worker bees. These cells serve multiple purposes: they are used for storing honey, pollen, and the queen's eggs. Honey is stored in the top part of the hive, as it is the colony's food source for the winter, while the lower sections are reserved for brood (eggs, larvae, and pupae).

Temperature regulation is also crucial inside the hive. Bees are incredibly sensitive to temperature and will work together to ensure their home remains between 93°F and 95°F, which is ideal for developing brood. If it gets too cold, bees will cluster together and vibrate their wings to generate heat. If it gets too hot, they'll fan the hive with their wings or bring in water to cool things down. It's a living system where each bee contributes to the hive's well-being, and understanding this dynamic will help you make better decisions as a beekeeper.

Pollination and Honey Production: The Bee's Magic

Perhaps the most magical part of beekeeping is witnessing the process of pollination and honey production firsthand. Bees play a critical role in pollinating plants, which is essential for the growth of many of the foods we rely on—fruits, vegetables, nuts, and more. As your bees forage for nectar, they transfer pollen from one flower to another, allowing plants to reproduce. It's not an exaggeration to say that without bees, our food supply would be in serious jeopardy.

When a worker bee collects nectar, she stores it in a special part of her body called the "honey stomach." Once she returns to the hive, she passes the nectar to other worker bees, who then reduce the water content by fanning their wings, transforming it into honey. The bees then seal the honey-filled cells with wax to preserve it for later use, especially during the winter when foraging is impossible.

The honey your bees produce is truly a miracle of nature—made from the nectar of countless flowers and the tireless work of thousands of bees. Each drop of honey represents hours of labor and an intricate process that's been refined over millions of years. As a beekeeper, your role is to support and protect this process, knowing that every jar of honey you harvest is a testament to the incredible world of bees.

CHAPTER 3

The Essential Gear You'll Need

When starting your beekeeping journey, it's easy to feel overwhelmed by the sheer number of tools, gadgets, and protective gear you're told you need. If you're anything like I was when I first began, you're probably asking yourself, "Do I really need all this?" The answer is, not necessarily. While certain equipment is non-negotiable for the safety of both you and your bees, there's also a lot of gear that you can grow into overtime. Let's break it down simply and practically, so you can start with what you truly need and understand why each piece of gear matters.

Beekeeping Suit and Protective Gear

One of the first things that may cross your mind when imagining yourself as a beekeeper is the iconic beekeeping suit. The thought of thousands of bees buzzing around you can understandably stir up feelings of nervousness, especially when you're just starting out. The good news is, modern protective gear makes beekeeping far less intimidating than it once was.

The beekeeping suit is your best friend when you're opening the hive, inspecting the bees, or harvesting honey. Most beginner beekeepers opt for a full-body suit that covers everything from head to ankle, with elastic cuffs around the wrists and ankles to prevent bees from sneaking inside. These suits are designed to be lightweight, allowing you to stay cool and comfortable while working in warmer weather, though no suit is completely sting-proof. It's a balance of protection and practicality, and over time, as you become more comfortable working with your bees, you may find that a simple jacket with a veil or just gloves are enough protection for you.

Speaking of veils, this is perhaps the most crucial part of your protective gear. Bees are naturally attracted to the head area, particularly around the eyes and mouth, so protecting your face is non-negotiable. The veil will prevent bees from getting too close to your face while still allowing you full visibility. Some beekeepers prefer a suit with an attached veil, while others use separate veils that can be worn with different clothing. It's all about finding what feels right for you.

Gloves are essential part of your gear, especially when you're starting out. They offer protection for your hands while you're handling the bees and hive equipment. However, gloves can sometimes limit your dexterity, making it harder to feel what you're doing when working with delicate bees or fragile comb. Some experienced beekeepers choose to forgo gloves for this reason, but in the beginning, I'd recommend wearing them. As you become more confident, you can decide whether or not to continue using them. The key is comfort—both your own comfort and ensuring that your bees feel gentle and respected in your hands.

Hives, Frames, and Tools: A Beginner's Guide

Once you're suited up, the next big step is setting up your hive. Choosing the right hive is a critical part of getting started, and there are several options out there, but most beginner beekeepers opt for the Langstroth hive. It's the most common hive design and the one you're most likely familiar with. It consists of stacked boxes, or "supers," which hold removable frames. These frames are where the bees build their comb, raise their young, and store honey.

The beauty of the Langstroth hive is that it's modular—you can start small and add more supers as your colony grows. This design also makes it easier to inspect your hive and harvest honey without disturbing the bees too much. The removable frames allow you to check for diseases, pests, and honey production efficiently; while also making sure your queen is laying eggs as she should. Other hive styles, such as top-bar or Warre hives, are available, but for a beginner, the Langstroth's straightforward design and ease of use make it a great choice.

When you first open your hive, you'll notice that each frame has a foundation. This foundation gives your bees a guide to start building their comb. Some beekeepers use wax foundations, which are a natural option, while others use plastic ones. The choice between these is largely personal and dependent on your goals. Natural beekeeping advocates might lean toward wax, as it's more aligned with what bees naturally use, but plastic foundations tend to be more durable and can last longer. Both have their pros and cons, but rest assured, your bees will build their comb beautifully on either.

Along with the hive and frames, you'll need some essential tools to manage your bees. The most important tool in your kit will likely be your hive tool—a small, flat instrument that helps you pry apart the frames and scrape off excess wax or propolis (the sticky resin that bees use to seal small gaps in the hive). Without a hive tool, separating the frames can be difficult, as bees tend to seal everything together with propolis.

indispensable tool is also the bee smoker. This simple device allows you to create a puff of cool smoke that calms the bees during hive inspections. When bees detect smoke, they instinctively retreat into the hive and gorge themselves on honey, preparing for what they believe could be a forest fire. In this state, they're much less likely to be defensive or sting, allowing you to work in peace. Using a smoker can feel a bit awkward at first—trying to keep the fire going while also managing your hive—but it's well worth the effort. It gives you the confidence to open the hive and inspect your bees without causing too much disruption.

You'll also need a bee brush, a soft-bristled tool that helps you gently move bees away from frames or other areas you need to inspect. The brush allows you to handle the bees with care, as harsh movements can harm them or cause agitation. Over time, you'll develop the patience to move slowly and deliberately, which is key to keeping both you and your bees calm.

Buying Your First Colony: What to Look For

Once you've got your gear and hive set up, it's time to bring the bee's home. This is where your adventure truly begins. But what should you look for when buying your first colony?

For beginners, I recommend purchasing a nucleus colony, often referred to as a "nuc." A nuc is essentially a small, established colony of bees that includes a queen, workers, drones, and some brood (baby bees). The benefit of starting with a nuc is that the bees are already familiar with one another, the queen is already laying eggs, and the colony is set up for success. It's like getting a head start, and it's generally less risky than buying a package of bees, which requires the bees to build their entire colony from scratch.

When purchasing your bees, make sure to find a reputable local beekeeper or apiary. Buying locally is important because your bees will already be adapted to the local climate and environment, giving them a better chance of thriving in your care. Ask the seller about the health of the colony and whether the queen is young and productive. A strong, healthy queen is the heart of a good colony, so don't be afraid to ask questions.

CHAPTER 4

Setting Up Your First Hive

Setting up your first beehive is one of the most exciting and nerve-wracking moments as a new beekeeper. There's a palpable sense of anticipation—your bees will soon arrive, and with them, the beginning of an incredible journey. I remember the feeling well: the thrill of knowing that soon, I would be part of this fascinating world, but also the anxiety of making sure everything was just right. If you're feeling the same mix of excitement and uncertainty, take a deep breath. This is where it all starts, and by preparing carefully, you'll set your bees—and yourself—up for success.

Choosing the Right Location

The first big decision you'll make is where to place your hive. Believe me, this is more important than it might seem at first. Bees are sensitive to their surroundings, and where you choose to set up their home will play a significant role in their productivity and health. It's not just about finding a spot that's convenient for you, but a place where the bees can thrive.

When I set up my first hive, I spent a lot of time wandering around my yard, trying to see things from a bee's perspective. You want your hive to be in a location that's sheltered from strong winds and gets plenty of sunlight, especially in the morning. Bees are solar-powered little creatures—they get to work as soon as the sun hits their hive, so having morning sun will encourage them to start their day earlier, which means more foraging and productivity.

Shade is also a factor to consider. While you want the hive to get some sun, too much heat can be stressful for the bees, especially in the hot summer months. A location that gets a mix of morning sun and afternoon shade is often ideal. You can also help regulate temperature with a hive stand, lifting the hive off the ground, which keeps it cooler and protects it from moisture.

Water is also a key consideration. Bees need water for cooling their hive and diluting honey to feed larvae. If you don't have a natural water source nearby, you can provide one by placing a shallow dish of water near the hive. Just be sure to add pebbles or floating objects so the bees have a safe landing place while they drink. It sounds simple, but these little details can make a big difference to your bees' comfort and efficiency.

Finally, think about accessibility. You'll be visiting your hive regularly, especially during the warmer months, so choose a spot that's easy for you to reach. You want to be able to check on your bees without having to struggle through tall grass or awkward terrain, but also somewhere slightly away from high-traffic areas of your yard, so the bees can come and go without interruption. Remember, bees tend to fly in a direct line out of the hive entrance, so positioning the entrance facing away from walkways or gathering spots will prevent accidental run-ins.

Assembling Your Hive: Step-by-Step

Now that you've chosen the perfect spot, it's time to put your hive together. This part can feel a bit daunting, but trust me—if you take it step by step, it's more straightforward than it seems. When I was assembling my first hive, I treated it almost like a puzzle. Each piece has its place, and seeing the hive come together is incredibly satisfying.

Start with the hive stand or a sturdy platform to elevate the hive off the ground. This simple step helps protect the hive from moisture and pests like ants and other insects. Plus, it makes hive inspections easier on your back!

Next comes the bottom board, which forms the foundation of your hive. Most beginners opt for a screened bottom board, which provides better ventilation and helps with pest control by allowing mites and debris to fall out of the hive rather than accumulate inside. Place the bottom board on the hive stand, ensuring it's level. A tilted hive can cause problems later on, with bees struggling to build their comb properly, so take a moment to check the alignment.

Once the bottom board is in place, it's time for the hive boxes, also known as supers. These will house your frames where the bees will build their comb, store honey, and raise their brood. Start with a deep super at the bottom—this will be your brood chamber, where the queen will lay her eggs. As the colony grows, you'll add more boxes on top, usually a mix of medium and shallow supers for honey storage.

The frames go inside the hive boxes, fitting snugly into place. You'll want to start with frames that have a foundation—this is a thin sheet of beeswax or plastic that gives the bees a guide for where to build their comb. It's helpful to keep the frames close together, as bees prefer narrow gaps (around 3/8 of an inch, to be exact). If the gaps are too wide, they might build excess comb in inconvenient places, making hive inspections more difficult.

The inner cover and outer cover go on top of the hive boxes. The inner cover provides some insulation and helps with ventilation, while the outer cover protects the hive from the elements. I remember being surprised at how crucial proper ventilation is—bees generate a lot of heat inside the hive, and poor airflow can lead to moisture buildup, which can be harmful to the bees, especially in the winter months.

Once everything is assembled, take a step back and admire your work. This is your bees' new home, and knowing that you've built it with care will make their arrival all the more meaningful.

Installing Your Bees: A Gentle Introduction

The day your bees arrive is unforgettable. There's a mix of excitement and a little nervousness, but mostly wonder. After all, you're about to introduce thousands of bees to the home you've lovingly prepared for them. Whether you're starting with a package of bees or a nucleus colony (a small, established colony with a queen and workers), the process of installing them into their new hive is one of the most important moments in beekeeping.

If you're using a package of bees, it may feel a bit surreal—after all, you'll be holding a box of around 10,000 bees! Take your time and work calmly. Bees are generally docile when they're without a hive, so while it's natural to feel a little apprehensive, remember that they're not out to sting you.

Begin by gently spraying the bees with sugar water. This helps calm them and gives them something to focus on besides flying around. Remove the queen cage from the package—it's a small, separate cage with the queen inside. She'll be the heart of your colony, so handle her gently. There's a candy plug at one end of her cage, and the bees will chew through this over the next few days to release her. This slow introduction ensures that the colony accepts her as their queen.

Place the queen cage in the center of the hive, wedged between two frames so she's secure but still accessible to the bees. Then, pour or shake the rest of the bees into the hive. Yes, it's a bit intimidating to have bees pouring out, but they'll quickly start clustering around the queen and settling into their new home.

For the next few days, leave the hive undisturbed except for checking that the queen has been released. Once she's free and laying eggs, the colony will begin to flourish. Watching them find their rhythm, flying in and out of the hive, foraging, and building comb, is one of the most rewarding sights in beekeeping.

Setting up your first hive is not just about logistics—it's about creating a space where your bees can thrive, and where you can learn and grow alongside them.

Every decision you make, from where to place the hive to how you handle the bees, plays a part in their success. And as your confidence grows, you'll look back on these early days with a sense of pride, knowing you've taken the first step in a lifelong journey with your bees.

CHAPTER 5

Caring for Your Bees

Caring for bees is a unique experience that deepens your connection to nature. It's not like tending to a garden or caring for a pet—it's about cultivating a relationship with a complex and delicate society of creatures. As a beekeeper, you become a steward, someone the bees rely on to create an environment where they can thrive, produce honey, and pollinate the surrounding ecosystem. This partnership is about nurturing their needs while they, in turn, nurture the land and provide you with that golden, sweet reward. But as much as bees are incredibly resilient and efficient, they also need your care to help them flourish, especially as the seasons change.

Feeding Your Bees: Nutrition and Health

Just like us, bees need a balanced diet to stay healthy. While they are excellent foragers and naturally gather nectar and pollen from the surrounding environment, there will be times when they need a little help from you. The food they collect from flowers is transformed into two primary sources of sustenance: honey and pollen. Honey provides them with carbohydrates, their main energy source, while pollen supplies the proteins, fats, vitamins, and minerals essential for their growth, especially for brood (young bees).

As a beekeeper, one of your first responsibilities is making sure your bees have enough to eat. In a perfect world, their environment would provide them with everything they need. However, in reality, there will be times when flowers are scarce, especially during late fall, winter, or early spring. This is where you come in. When the natural food supply is low, supplementing their diet can make the difference between a thriving hive and one that struggles to survive.

During the spring and summer, when flowers are abundant, bees are busy collecting nectar and pollen, so you won't need to interfere much. But during the fall, as they begin preparing for winter, it's crucial to monitor their food stores. If the hive hasn't gathered enough honey to sustain them through the colder months, you'll need to provide supplemental feeding, usually in the form of sugar syrup. A simple mixture of sugar and water mimics the energy bees get from nectar and helps them survive when flowers aren't blooming. In addition, some beekeepers use pollen substitutes or supplements to ensure the bees are getting enough protein, especially if they've been foraging in an area with limited floral diversity.

It's important to remember that bees are not just honey-making machines— they're living creatures with intricate needs. Keeping an eye on their nutrition is one way to help them maintain their health. A colony that's well-fed will be more resilient to diseases, pests, and the stresses that come with changing seasons.

Seasonal Care: Spring, Summer, Fall, and Winter

Caring for bees is a year-round responsibility, and the way you care for them changes with each season. Just as the bees shift their activities with the seasons, so too must your approach to beekeeping. Knowing what your bees need throughout the year will help you plan ahead and ensure their survival.

Spring is a time of renewal in the hive. As the days grow warmer, your bees become more active, foraging for nectar and pollen to kick-start the hive's expansion. The queen will begin laying more eggs to build up the colony, and you'll see the hive come to life after the winter's dormancy. Your role in spring is to help them manage this burst of activity. This is when you'll inspect the hive regularly to check for overcrowding. A thriving hive can quickly run out of space, and if the bees feel cramped, they may swarm—meaning a portion of the colony will leave to start a new hive elsewhere. While swarming is a natural process, it's something you want to manage, especially as a beginner, so you don't lose half your bees. Adding new hive boxes and giving the bees more room can help prevent this.

Summer is when the bees are at their busiest. The hive is buzzing with activity as the foragers fly out in search of nectar and pollen, and the workers inside are tending to the brood and building comb.

Your job during the summer is mostly to monitor the hive and ensure they have enough space for all the honey they're producing. It's also a good time to harvest honey, but be mindful not to take too much. The bees need plenty of honey stored away to survive the winter. In addition, summer is the time to stay vigilant about pests and diseases. Varroa mites, one of the most destructive parasites for bees, can become a serious problem if left unchecked. Regular hive inspections and treatments, if necessary, can help keep your bees healthy through the summer months.

As **fall** approaches, the bees start preparing for the cold months ahead. The queen will slow down her egg-laying, and the workers will focus on storing up enough honey to last the winter. This is the time for you to assess the hive's food stores. If it looks like they haven't collected enough honey, you'll need to begin supplemental feeding. In addition, you should start preparing the hive for winter by reducing the entrance (to help keep out drafts and unwanted pests) and making sure the hive is insulated enough to withstand the cold. In some climates, beekeepers even wrap their hives in insulation to protect them from harsh winter temperatures.

Winter is when your bees go into a sort of hibernation. They cluster together in the hive, vibrating their wings to generate warmth, and survive on the honey they've stored. During this time, your bees won't be flying or foraging, and you won't need to open the hive unless absolutely necessary. Winter can be a quiet time for beekeepers, but it's also a time of uncertainty, as you hope that your colony has enough resources to make it through until spring. Regularly checking on the hive's food stores, and providing emergency feeding if needed, is crucial to ensuring your bees survive the winter.

Monitoring the Hive: What to Watch For

Caring for your bees means keeping a close eye on the health and activity of the hive. Regular inspections are one of the most important parts of beekeeping, and they allow you to detect potential problems early on before they become serious issues. But what should you be looking for?

First, always check on your **queen**. She's the heart of the colony, and without her, the hive will eventually die. During inspections, make sure the queen is present and laying eggs. If you can't find her, look for signs of her presence—like eggs and larvae. If the queen isn't laying, or if the colony appears queenless, you may need to introduce a new queen to the hive.

You'll also want to monitor the hive's **brood pattern**. A healthy brood pattern should be consistent, with eggs, larvae, and capped cells arranged in a neat, compact area. Spotty brood patterns, where cells are scattered or missing, could indicate that the queen is failing, or that there's a disease affecting the hive.

Next, keep an eye out for **pests and diseases**. Varroa mites are one of the biggest threats to honeybees, and their presence can quickly weaken a colony. Other issues to watch for include wax moths, small hive beetles, and foulbrood disease. Regular monitoring will help you catch these problems early and treat them before they cause too much harm.

Observe the **general activity** of the bees. Are they flying in and out of the hive regularly? Are they bringing in pollen? A healthy hive will be bustling with activity during the spring and summer. If the bees seem lethargic or inactive, it could be a sign of trouble.

CHAPTER 6

Hive Inspections: What to Expect

Opening up a beehive for inspection can feel like opening a treasure chest. Each time you lift that lid, you're stepping into a hidden world—one filled with thousands of tiny lives, all working in unison for the survival and success of their colony. As a beekeeper, inspecting your hive is one of the most important tasks you'll do. It's your window into the health and well-being of your bees, and it's where you'll learn to read the subtle signs of how they're doing. At first, these inspections can feel a bit daunting. You might be worried about disrupting the hive or unsure of exactly what you're looking for. But with time and experience, each inspection will become second nature. It's not just about checking boxes or following steps—it's about building a relationship with your bees and understanding what they need to thrive.

How Often Should You Inspect?

One of the first questions new beekeepers often ask is, "How often should I inspect my hive?" It's a great question because there's a balance to strike here. You want to be attentive and proactive, but you also don't want to disturb your bees too much. Each time you open the hive, you interrupt their work, their rhythm. So, inspections should be frequent enough to catch any issues early, but not so often that you stress the bees out.

In the spring and summer, when the colony is most active, you'll likely want to inspect the hive about every 7 to 10 days. This is the period of growth and expansion. The queen is laying eggs, and the workers are busy building comb, raising brood, and collecting nectar and pollen. During this time, regular inspections are important to monitor the health of the queen, the development of the brood, and the progress of honey storage. It's also when you'll be on the lookout for signs of swarming, which can happen when the colony becomes overcrowded.

As you move into the fall, the frequency of inspections can taper off. The bees are preparing for winter, and there's less need to open the hive. A check every few weeks should suffice to ensure they have enough food and that the hive is ready for the cold months ahead. Once winter arrives, inspections become minimal. You don't want to disrupt the cluster of bees huddling for warmth, so only open the hive if absolutely necessary—such as to check food stores or in the case of an emergency.

It's important to note that while regular inspections are key, every hive and every season is different. Some hives may need more frequent attention, especially if there's a particular issue or if it's a weak colony. And sometimes, simply observing the hive from the outside can tell you a lot. Are bees flying in and out? Are they bringing in pollen? Are there dead bees piling up near the entrance? These observations can help guide you in deciding when it's time to open up the hive.

Signs of a Healthy Hive

So, what are you looking for when you inspect a hive? A healthy colony will show certain signs that everything is running smoothly. Understanding these signs is key to knowing that your bees are thriving.

The first thing you'll want to check is the presence and health of the queen. The queen is the heart of the hive—without her, the colony cannot survive. During an inspection, you may not always see her, and that's okay. What's more important is that you see signs of her presence, like fresh eggs or larvae. If you do spot the queen, she should appear large, with a long abdomen. Her behavior should be calm and focused, as she moves from cell to cell laying eggs. A good queen will lay eggs in a neat and consistent pattern, with very few gaps between them. This is called a solid brood pattern, and it's one of the clearest indicators of a healthy queen and a well-functioning colony.

Next, take a look at the brood (the developing bees). A strong, healthy hive will have a mix of eggs, larvae, and capped brood (those that are nearing the end of their development). The brood pattern should be tight and continuous, with only a few empty cells. Spotty brood patterns—where there are many gaps or inconsistencies—can be a sign of a failing queen or a disease affecting the brood.

Healthy bees themselves will appear active and industrious. They'll be buzzing around the hive, moving from frame to frame, building comb, storing honey, and feeding the brood. You'll notice some bees with yellow or orange pollen baskets on their legs as they return from foraging. This is a good sign that they're collecting food and supporting the hive's growth.

You are to check if there is honey production. You'll see honey stored in the upper corners of the frames, with the bees capping it off with a thin layer of wax when it's ready. Healthy bees will keep these stores well-stocked, especially as they prepare for winter. Beeswax itself is also a good indicator. Clean, white wax is a sign that your bees are actively building and expanding, while darker, older wax means the comb has been used for brood rearing or honey storage over multiple seasons.

A healthy hive will also smell good—there's a sweet, earthy scent that comes from the mix of beeswax, honey, and the bees themselves. If you notice any foul or off-putting odors, it could be a sign of disease or decay within the hive.

Common Problems and How to Fix Them

Unfortunately, not every inspection will go perfectly, and that's okay. Beekeeping, like anything in nature, comes with challenges. But by knowing what to look for and how to respond, you can often fix problems before they become serious.

One common issue you may encounter is a failing or absent queen. If your hive has no eggs, or if the brood pattern is spotty and inconsistent, it could mean the queen isn't laying properly or that she's gone altogether. If this happens, you'll need to replace her. Some beekeepers will introduce a new queen, which can be purchased from a local breeder. The process of introducing a new queen requires some care, as the colony needs time to accept her. But with patience, a new queen can restore the hive's health.

Also watch out for this problem called **varroa mites**—tiny parasites that can wreak havoc on a colony if left unchecked. During inspections, you might notice these mites on the bees themselves or on the larvae. There are several methods to manage varroa mites, from chemical treatments to more natural approaches, like screened bottom boards or powdered sugar dusting. Regular mite checks, especially during the late summer and early fall, are essential to keeping this problem under control.

You might also encounter **foulbrood**, a bacterial disease that affects the brood.

It's one of the more serious diseases a hive can face. Signs of foulbrood include sunken, discolored brood cells and a rotten smell. If you suspect foulbrood, it's important to act quickly, as it can spread to other hives. In some cases, the best course of action is to destroy the infected frames or even the entire colony to prevent contamination.

Swarming can also be a challenge, especially during the spring and early summer. If your hive is overcrowded, the bees may decide to swarm, which means the queen and a large portion of the workers leave to start a new colony. While swarming is natural, it's something you want to manage. By giving the bees more space, splitting the hive, or cutting back on the queen's egg-laying, you can reduce the likelihood of swarming.

CHAPTER 7

The Most Common Threats to Your Bees

If you're new to beekeeping, you might be surprised at just how many threats there are to your hive. The most common ones include Varroa mites, small hive beetles, wax moths, and various bacterial or fungal infections. Each one presents unique challenges, and they all require a watchful eye and proactive management.

Varroa mites are often called the "bane of beekeeping." These tiny, crab-like parasites attach themselves to bees and feed on their bodily fluids, weakening them and transmitting deadly viruses. Even a small population of Varroa mites can severely impact your colony's health. These mites are highly invasive and can easily spread from hive to hive, making them a significant concern for beekeepers everywhere.

Small hive beetles are common pest, especially in warmer climates. These beetles lay their eggs in the hive, and once the larvae hatch, they feed on honey and pollen, leaving behind a fermented, slimy mess that's harmful to your bees and honey production.

Wax moths are also a concern. They lay eggs inside the hive, and their larvae burrow through honeycombs, feeding on the wax and causing damage to the structure of the hive itself. While they typically target weaker colonies, any hive can be vulnerable if not carefully monitored.

Then there are the diseases: **American foulbrood**, **Nosema**, and **Chalkbrood** are a few that can devastate a hive if not treated promptly. These bacterial and fungal infections often attack the brood, the bee larvae, weakening the colony's ability to replace older bees.

For a new beekeeper, encountering any of these problems can feel like a personal failure. But it's important to understand that pests and diseases are a natural part of beekeeping, and even the healthiest, most well-cared-for hives can fall prey to them. The key is early detection and management.

Natural and Effective Ways to Protect Your Hive

Prevention is always better than cure, especially when it comes to beekeeping. There are several natural and effective methods to protect your hive from pests and diseases, many of which center around creating an environment where your bees can thrive and pests have fewer opportunities to take hold.

First, maintaining strong, healthy colonies is one of the best defenses against both pests and disease. Strong colonies are more capable of defending themselves against intruders like small hive beetles and wax moths, and they're better equipped to survive an outbreak of disease. Keeping your bees well-fed and providing a safe, clean environment goes a long way toward ensuring their strength.

You can also implement natural methods to manage pests like Varroa mites. For instance, many beekeepers use a screened bottom board in their hive, which allows mites to fall through the floor rather than reattaching to the bees. Powdered sugar dusting is also non-chemical approach. By dusting your bees with powdered sugar, you cause the mites to slip off the bees as they clean themselves. Additionally, breeding and using bees that have developed natural resistance to Varroa mites is becoming more common. These bees, often called "hygienic bees," are better at grooming themselves and removing mites from the hive.

When it comes to pests like small hive beetles, simple mechanical traps can be placed inside the hive. These traps usually contain oil or a liquid that drowns the beetles, and they're an easy, chemical-free way to keep the beetle population in check. Try to also keep your apiary clean and free from excess debris, as beetles are often attracted to decaying matter. For wax moths, freezing old comb before storing it can kill any larvae and prevent future infestations.

It's also vital to avoid overloading your bees with chemicals. While some chemicals may be effective in controlling pests, they can also harm your bees or contaminate your honey. Wherever possible, choose natural methods first, and only resort to chemical treatments

when absolutely necessary. Your bees
will be healthier and happier for it, and
you'll enjoy cleaner, more natural honey.

When to Seek Professional Help

Even with the best of care, there will be
times when your hive faces threats that
you can't handle on your own.
Recognizing when to seek help is a
critical part of being a responsible
beekeeper. Sometimes, even the most
experienced beekeeper finds themselves
dealing with a particularly aggressive
pest infestation or an outbreak of disease
that requires more specialized knowledge
or resources.

For example, if your hive shows signs of
American foulbrood, you should
contact your local beekeeping
association or an experienced beekeeper
immediately. This bacterial infection can
wipe out entire colonies, and in many

cases, infected hives need to be destroyed to prevent the spread to other colonies. Many regions require beekeepers to report cases of foulbrood, so it's important to act quickly to contain the problem.

In cases of severe **Varroa mite infestations**, where natural methods haven't been effective, professional advice may be necessary to help you navigate more advanced treatments. There are chemical treatments available, but they need to be used correctly to avoid harming the bees. An experienced beekeeper or apiary inspector can guide you in choosing and applying these treatments safely.

Similarly, if your bees are being overwhelmed by pests like **small hive beetles** or **wax moths**, it may be time to call in help. These pests can sometimes multiply faster than a beginner can manage, especially if they've gained a foothold in your apiary. A mentor or local beekeeping group can often provide hands-on assistance, helping you save your colony before the problem becomes too severe.

Beekeeping is a journey, and like any journey, it comes with its share of bumps in the road. But with patience, attention, and a willingness to seek help when needed, you'll find that the rewards far outweigh the challenges. Your bees rely on you to keep their home safe and healthy, and by being vigilant about pests and diseases, you're ensuring that your colony will thrive for many years to come.

CHAPTER 8

Harvesting Honey

Harvesting honey is one of the most rewarding moments in a beekeeper's journey. After months of caring for your bees, watching them build their hive, and providing them with the resources they need to thrive, you finally reach the stage where you can collect the golden treasure they've created. However, honey harvesting isn't just about reaping the rewards. It requires patience, care, and respect for your bees. Whether you're a beginner or have a few seasons under your belt, understanding when and how to harvest honey safely and ethically is key to building a sustainable and successful beekeeping practice.

Knowing When Your Honey is Ready

One of the first things you'll need to learn as a beekeeper is how to tell when your honey is ready for harvesting. This is not just a matter of timing or gut feeling — it's about knowing what to look for within your hive. Bees work hard to gather nectar, and it takes time for them to convert it into honey. The nectar they collect is initially too watery, and the bees will fan it with their wings to reduce its moisture content. Only when the honey reaches the proper moisture level (usually around 17-18%) will the bees seal it in the honeycomb with a thin layer of wax. This process is known as "capping."

The key indicator that your honey is ready for harvesting is when a significant portion of the honey cells are capped. The capping is like nature's way of saying, "This honey is done!" If you try to harvest too early, when the honey hasn't been properly capped, it may still contain too much moisture. This can cause the honey to ferment, which not only affects the taste but also diminishes its shelf life.

Patience is crucial during this stage. You might feel eager to harvest, especially if you see your bees working busily and filling up the frames, but taking honey before it's ready can do more harm than good. Give your bees time to finish their process — you'll be rewarded with high-quality, delicious honey that will last.

How to Safely and Ethically Harvest Honey

When it comes to harvesting honey, it's important to approach the process with care, both for your bees and for the quality of the honey itself. The first step is to make sure you have the right tools: a honey extractor, bee brush, uncapping knife, and protective gear. You'll also need a calm, clear day to work, as this reduces stress on the bees and makes the process smoother.

Before you begin harvesting, it's a good idea to do a final hive inspection. This allows you to check the health of the colony and ensure that there's enough honey left for the bees. Remember, you're not taking *all* of their hard work — the bees rely on honey as their primary food source, especially during the colder months. As a general rule, always leave enough honey in the hive to sustain the colony through winter. Ethical beekeeping means prioritizing the health of your bees over the amount of honey you collect.

To safely harvest the honey, you'll first need to remove the bees from the honey frames. One gentle method is to use a bee escape, which encourages bees to leave the honey supers without harming them. Alternatively, you can lightly brush the bees off the frames using a bee brush, being mindful not to damage the honeycomb or stress the bees.

Once the bees are cleared from the frames, you'll need to uncap the honeycomb. This involves gently slicing off the wax capping that the bees placed over the finished honey. An uncapping knife or fork works well for this. Be sure to save the wax capping, as they can be melted down and used for other purposes, like making candles or lip balm.

The next step is extracting the honey. Using a honey extractor, place the uncapped frames inside the machine, which spins the frames and forces the honey out through centrifugal force. This is a much more efficient and bee-friendly method than crushing the honeycomb, as it preserves the comb for future use. Once the honey is extracted, it's passed through a fine mesh filter to remove any bits of wax or debris, leaving you with pure, clean honey.

Throughout the process, maintain a gentle touch. Bees are sensitive creatures, and their hive is their sanctuary. Ethical harvesting means taking only what you need and allowing the bees to continue their work without disruption. Remember, the bees have worked tirelessly to create this honey, and they deserve your respect. By treating your bees with care, you not only ensure a sustainable honey harvest but also foster a healthier, happier colony.

Storing and Using Your Harvest

Now that you've harvested your honey, you'll want to store it properly to maintain its freshness and flavor. Honey is unique in that it never truly spoils, thanks to its low moisture content and natural antibacterial properties. However, improper storage can lead to crystallization or changes in taste over time.

The best way to store honey is in a sealed, airtight container. Glass jars are a popular choice because they don't affect the flavor of the honey, and they provide a good seal to keep moisture out. Honey should be stored at room temperature in a cool, dark place, like a pantry. While honey doesn't need refrigeration, extreme heat or cold can cause it to change consistency. If your honey does crystallize, don't worry — this is a natural process that happens to raw honey. Simply place the jar in warm water and stir gently until the crystals dissolve.

Once your honey is stored, the fun begins! The uses for honey are endless. It can be enjoyed straight from the jar, drizzled over toast, stirred into tea, or used in baking. Honey's versatility makes it a staple in any kitchen. You can also give honey as gifts, sharing the fruits of your labor with family and friends.

But honey's uses extend beyond the kitchen. It has long been valued for its medicinal properties. Raw honey contains antioxidants, enzymes, and natural anti-inflammatory compounds, making it an excellent remedy for sore throats, coughs, and skin irritations. You can create homemade skincare products, like honey face masks or scrubs, which are gentle and nourishing for the skin.

One of the most fulfilling aspects of harvesting honey is knowing that you're using a natural, sustainable product that you helped create. The bond you form with your bees throughout the process makes that jar of honey all the more special. It's not just about having something sweet to spread on your toast — it's about being part of a symbiotic relationship with nature.

As you grow as a beekeeper, you'll find that each harvest feels like a milestone. There's a sense of accomplishment in knowing you've provided a safe, thriving environment for your bees while also reaping the benefits of their hard work. Harvesting honey is more than just an end result; it's a testament to the care and respect you've shown for your bees, and that respect will continue to reward you year after year.

CHAPTER 9

How Beekeeping Helps the Environment

Bees are often referred to as "nature's pollinators," and for good reason. Pollination is a critical process in the reproduction of plants, and bees are responsible for pollinating approximately 75% of the world's flowering plants, including about one-third of the crops we consume. Without bees, the availability and diversity of fresh produce would significantly decline, affecting everything from fruits and vegetables to nuts and seeds. By keeping bees, you're directly contributing to the pollination process and helping sustain the plant life that so many creatures, including humans, rely on.

But bees don't just help pollinate the crops in your local area. When you set up a beehive, you're creating a ripple effect that extends far beyond your own backyard. Bees typically travel up to five miles from their hive in search of nectar, pollinating a wide variety of plants along the way. This means your bees are not only benefiting your garden, but also helping nearby farms, wildflowers, and natural habitats. Over time, a healthy bee population can significantly improve the biodiversity of an area, leading to stronger ecosystems that support a broader range of wildlife.

In a world where pesticide use, habitat loss, and climate change are all taking a toll on bees, beekeeping provides a much-needed sanctuary for these vital pollinators. By giving bees a safe and stable environment to thrive, you're helping to combat the decline of bee populations and playing a part in the global effort to protect biodiversity.

Supporting the Bee Population

In recent years, the plight of bees has become a growing concern. Colony collapse disorder, pesticide exposure, and habitat destruction have all contributed to a dramatic decline in bee populations worldwide. As a beekeeper, you have a unique opportunity to directly support the health and vitality of bees, not just in your own hives, but in the wider ecosystem as well.

One of the most important ways you can support the bee population is by fostering strong, healthy colonies. This starts with providing bees with the right environment. Ensure that your bees have access to a diverse range of plants, flowers, and trees that provide nectar and pollen throughout the seasons. If you live in an urban or suburban area, planting bee-friendly flowers in your garden or encouraging neighbors to do the same can help create an oasis of resources for your bees. Wildflowers, clover, lavender, and native plants are excellent choices that bees love.

Additionally, reducing the use of pesticides in your garden and surrounding areas is crucial for the health of your bees. Pesticides, especially neonicotinoids, are harmful to bees and can cause disorientation, making it difficult for them to return to the hive. Even if you're managing your own space responsibly, be mindful of what's happening in nearby areas, and where possible, advocate for more bee-friendly practices in your community. Every small step you take to reduce chemical exposure for your bees is a step toward improving their survival and health.

If you're looking to expand your impact, consider working with local organizations or conservation groups that focus on pollinator protection. Many regions have programs dedicated to restoring native bee populations or promoting pollinator-friendly practices.

Whether through education, planting initiatives, or policy advocacy, joining these efforts can amplify the positive effect you're having as a beekeeper.

Tips for Ethical Beekeeping Practices

Beekeeping is a delicate balance between reaping the rewards of honey and maintaining the health and happiness of your bees. Ethical beekeeping is about making decisions that prioritize the well-being of the bees above all else. While the honey is a beautiful byproduct of their hard work, it's important to remember that bees are not simply "honey producers." They are complex, intelligent creatures with their own needs and instincts, and it's our responsibility as beekeepers to respect that.

One of the key principles of ethical beekeeping is ensuring that your bees have enough honey to sustain themselves, especially during the winter months. While it may be tempting to harvest as much honey as possible, ethical beekeepers understand that bees need a substantial reserve of honey to survive when nectar is scarce. Always leave enough honey for your colony to thrive. Many experienced beekeepers recommend leaving at least 40 to 60 pounds of honey in the hive, depending on your region and climate.

Ethical beekeeping also means paying close attention to the health of your hive. Regular inspections are essential, not just to monitor honey production but to check for signs of disease, pests, or other stressors that could harm your bees. Keeping your hive clean, providing adequate ventilation, and managing pests like Varroa mites or small hive beetles in a way that doesn't rely heavily on chemicals will contribute to a more sustainable beekeeping practice. Opting for natural treatments and solutions, whenever possible, helps maintain the health of the bees without introducing harmful substances into the hive.

An important aspect of ethical beekeeping is avoiding over-intervention. Bees have survived for millions of years without human interference, and sometimes the best thing you can do is allow them to follow their natural instincts. While it's important to provide them with a safe environment, over-managing the hive can lead to unnecessary stress for the bees. Finding a balance between careful management and letting nature take its course is one of the most rewarding aspects of ethical beekeeping.

Education and community involvement play a crucial role in ethical beekeeping. Sharing your knowledge with others, whether through local beekeeping groups, workshops, or even conversations with neighbors, helps foster a culture of respect for bees and other pollinators. By spreading awareness about the importance of bees and how to protect them, you can inspire others to make bee-friendly choices in their own lives, creating a ripple effect of positive change.

Beekeeping, when done ethically, is a practice rooted in harmony with nature. It's about fostering a relationship with the bees, caring for them, and giving them what they need to thrive. In return, they give us not only honey, but a healthier, more vibrant environment. By prioritizing sustainability and ethical practices, you're not just helping your bees — you're contributing to a brighter future for the planet. The work you do as a beekeeper is part of a larger movement to protect our pollinators, and every jar of honey you harvest is a testament to your care, respect, and commitment to the natural world.

CHAPTER 10

What if My Hive Fails?

It's every new beekeeper's fear: you invest your time, money, and energy into setting up a hive, only to watch it collapse. Hive failure can happen for several reasons — pests, diseases, harsh weather conditions, or even queen failure. When it happens, it can feel disheartening, as though you've done something wrong. The first thing I want to tell you is this: don't blame yourself. Hive failure is not uncommon, even among seasoned beekeepers. Bees are delicate creatures, and there are many factors that can influence their success, some of which are beyond your control.

The most important step to take when faced with hive failure is to approach it as a learning experience. Instead of seeing it as a loss, see it as an opportunity to understand what went wrong. Did the hive fall victim to Varroa mites? Did the queen die, leaving the colony leaderless? Were there signs of poor nutrition or a lack of forage in the area? By conducting a thorough inspection and reflecting on your beekeeping practices, you can identify potential causes and take steps to prevent the same issue from occurring in the future.

It's also helpful to connect with more experienced beekeepers when facing hive failure. Joining a local beekeeping group or forum can give you access to a wealth of knowledge and support. Often, seasoned beekeepers have faced similar challenges and can provide valuable insights or solutions that you might not have considered. Beekeeping is as much a community endeavor as it is an individual one, and there's no shame in asking for help.

Most importantly, don't give up. It's easy to feel defeated when a hive doesn't make it through the season, but remember that every failure is a stepping stone toward success. Beekeeping is a long-term commitment, and setbacks are part of the journey. With each challenge, you grow as a beekeeper, and the lessons you learn will make you more resilient and better equipped to care for future hives.

Dealing with Bee Stings and Fear

Let's face it — the idea of being stung by a bee can be a source of anxiety for many beginner beekeepers. After all, bees have stingers for a reason, and getting stung is a possibility every time you work with your hive. However, the fear of bee stings shouldn't stop you from enjoying the beauty and rewards of beekeeping. With time, knowledge, and experience, you'll find that you can manage both the physical discomfort of stings and the emotional fear that might come with them.

First, it's important to understand that bees don't sting without reason. Honeybees are generally non-aggressive and will only sting when they feel threatened or when their hive is in danger. When you approach your hive calmly and respectfully, wearing the proper protective gear, the risk of being stung decreases significantly. Over time, you'll develop a sense of how your bees behave and when they're more likely to become defensive, allowing you to work with them in a way that minimizes the chance of stings.

But what happens if you do get stung? It's important to stay calm. For most people, a bee sting causes temporary discomfort, including redness, swelling, and mild pain. However, if you have an allergy to bee stings, it's crucial to carry an epinephrine injector (EpiPen) and seek medical attention if a sting occurs. For non-allergic reactions, the best course of action is to remove the stinger as quickly as possible (you can scrape it off with a fingernail or a credit card), wash the area with soap and water, and apply a cold compress to reduce swelling.

If you find that fear of being stung is holding you back from fully engaging with your bees, it can be helpful to take small, gradual steps toward building confidence. Start by spending time near your hive without opening it, simply observing the bees' flight patterns and behavior. This allows you to become more comfortable in their presence. Over time, as you become more accustomed to working with your bees, the fear of stings will likely diminish. Keep in mind that every beekeeper, even the most experienced ones, gets stung occasionally. It's a natural part of the process, but it doesn't define your beekeeping experience.

Overcoming Setbacks and Staying Motivated

In beekeeping, as in life, setbacks are inevitable. Whether it's a hive that didn't survive the winter, a swarm you couldn't catch, or a batch of honey that didn't turn out as expected, you'll face moments of frustration. The key to staying motivated during these times is to keep the big picture in mind. Beekeeping is a journey, and like any worthwhile endeavor, it comes with its ups and downs.

One of the best ways to maintain motivation is to remind yourself why you started beekeeping in the first place. Maybe it was the desire to connect with nature, to produce your own honey, or to contribute to the health of your local environment. Whatever your reason, hold onto that sense of purpose when things get tough. Reflect on the joy you felt when you first set up your hive or when you spotted your bees happily buzzing around your garden. These moments of joy are what sustain you through the challenges.

It's also important to celebrate your successes, no matter how small. Did you successfully complete your first hive inspection? Did you spot the queen? Did your bees survive their first winter? Each of these achievements is a testament to your growth as a beekeeper, and acknowledging them helps build the resilience needed to face future challenges.

If you find yourself feeling discouraged, take a break if necessary. Stepping back for a moment doesn't mean you're giving up — it's a way to recharge and come back with a fresh perspective. During this time, consider seeking inspiration from other beekeepers. Whether it's reading a book, watching a documentary, or attending a local beekeeping class, learning from others can reignite your passion and offer new ideas for overcoming challenges.

Ultimately, beekeeping is about building a relationship with your bees, with nature, and with yourself. The setbacks you face are opportunities for growth and learning, and each one brings you closer to becoming a more knowledgeable and empathetic beekeeper. Remember that even the most successful beekeepers started where you are now, and they, too, had to navigate the same challenges. Stay patient, stay curious, and most importantly, stay motivated — your bees, and the rewards of beekeeping, are worth it.